RUNNING
ON EMPTY

Origins of the Universe

Gilberto Davila Jr

Dedication

For my children, Matthew and Emma

Acknowledgment

I thank my loving wife, Tatyana, for her patience, support and understanding through this long journey.

I thank my parents and other family members for their support and guidance, who have always encouraged me to never give up in seeking the truth.

I wish to express my sincere appreciation to a long time friend and mentor, Honorable Donald H. Birnbaum for his guidance and direction. His vast knowledge and criticism greatly contributed to the completion of this book.

A special thanks to a wonderful teacher and spiritual advisor, Kadam JoAnn Lawrence along with other members of Samudrabadra Kadampa Buddhist Center.

I would also like to express my gratitude to Dr. George Naryshkin, Dr. David Berko and Paul Beaulieu and many other individuals who provided support, great intellectual insight and comments.

I like to thank the following organizations that provided me with an opportunity to study and learn about our universe. Without these facilities, educational programs, books and websites, it would not have been possible to complete my studies and combine various schools of thought that lead me to complete this book.

The Museum of Natural History, New York, NY
The Haden Planetarium, New York, NY
The National Geographic organization
NASA Facilities and Centers
Coral Castle, Homestead, FL
The Edison and Ford Museum, Fort Myers, FL
The Discovery Channel
Channel 13, New York, NY
New York Public Library

PBS.org
CERN.ch
Popular Science
Mkaku.org
PhysOrg.com
Space.com
The following publications were referenced in the research and completion of this book.

Albert Einstein. *Ideas and Opinions.* New York, NY: Random House, 1988.

Albert Einstein, Robert W. Lawson. *Relativity: The Special and the General Theory.* Berkeley, CA: Pearson Education Inc, 2007

Albert Einstein, Hermann Minkowski. *The Principle of Relativity; original Papers.* University of Toronto; University of Calcutta: 2010

Daniel C. Mattis. *The Theory Of Magnetism Made Simple: An Introduction To Physical Concepts And To Some Useful Mathematical Methods.* Hackensack, NJ: World Scientific, 2006.

Edward Leedskalnin. *Magnetic Current, Mineral, Vegetable and Animal Life.*Homestead, Florida: Rock Gate, 1945. Mokelumne Hill, CA: Mokelumne Press, 1988, 1945.

Gyatso, Geshe Kelsang. *Joyful Path of Good Fortune.* Glen Spey, NY: Tharpa Publications, 2006.

James D. Livingston. *Driving Force: The Natural Magic of Magnets.* Cambridge, MA: Harvard University Press, 1996.

Michio Kaku. *Physics of The Impossible: A Scientific Exploration Into The World Phasers, Force Fields, Teleportation and Time Travel.* New York, NY: Anchor Books: Division of Random House, 2009.

Nikola Tesla; Edited by Thomas Commerford Martin. *The Inventions, Researches and Writings of Nikola Tesla.* Los Angeles, CA: Angriff Press, 1981.

Nikola Tesla, Ben Johnston, Editor. *My Inventions: The Autobiography of Nikola Tesla.* Austin, Texas: Hart Brothers, 1982

Percy Seymour. *Cosmic Magnetism.* Bristol, UK: Adam Hilgar, 1988

Thomas A. Edison, Professor Paul B. Israel, Editor, Professor Keith Nier, Editor. *The Papers of Thomas A. Edison: From Workshop to Laboratory, June 1873-March 1876.* Baltimore, MD: The Johns Hopkins University Press, 1992.

Walt Whitman. *Leaves of Grass: The Original 1855 Edition.* Mineola, NY: Dover Publications, 2007.

Content

Introduction

I rise to another day and my mind is awakened. I am looking forward to what today will bring but expecting it will be no different than any other. My search for answers has driven me to more questions and the questions have drawn me closer to more answers. Is this the path I am supposed to take? Yes, it is, but why?

What is life and what is the meaning of life? There has to be something bigger out there. I ask myself, how was universal life created, and what were the factors behind it? So many questions have made me ask myself, was everything in our universe dependent upon itself? Is life just born? Is it "given," and can it truly be created from nothing? Am I—and all mankind—independent from any and all other existing forms of life? Does humanity exist freely, independent from all other phenomena of life?

I am in absolute awe that answers actually exist. I have been afforded the gift of universal knowledge: the correlation of the universe, galaxies, stars, atoms, protons, neutrons, the cosmic North and South polarities, life, love, hate, greed, fear and ignorance, animals, insects, trees, plants, flowers, grass, mountains, rocks, sand, dust, the emptiness of time and space where a singularity joins with a plurality to create all.

The search for answers has delivered me to freedom. The understanding of universal physical laws is undeniable fact. These have been and will always be right in front of us all. Fundamental explanations have given rise to the understanding of once inexplicable phenomena. We must accept dispassionately whatever we learn without fear of consequences. My desire no longer needs fueling. All that was, all that is, and all that will be has been delivered to me. I have found a profound, unimaginable liberty of thought and, therefore, life.

What many others may believe to be unanswerable, untrue, or pointless is of no concern to me. These words are mine as much as these words are yours; these words are of the pure universe. I have passed countless decades on a quest to find the origins of life itself: wondering, wanting, praying, begging, and wishing for answers. At first I believed that a personal quest was driving me, but it was not mine at all. It was a quest on behalf of humanity. My search was not born out of idle curiosity, because every path I took had a purpose, and my search has now ended.

Let's begin.

The universe's origins are truly some of the most puzzling questions for mankind. Along with the concepts of time and space, science has striven to answer many questions about how the universe and humankind were created. There is so much to learn about the universe. What we understand now is miniscule compared to what there is to learn, but it is as significant as a single grain of sand in a desert. One grain of sand may appear to be insignificant, but when added to great numbers of similar materials, vast deserts are created. These grains of sand are like the beginning of what exists and needs to be understood.

To date no one is really sure how, where, or why things occur. In some areas of the scientific community, several questions are answered while many others are mired in conflicting theories. The quest for scientific answers and the thirst for knowledge are gifts to humanity.

I, according to my own unconventional standards, am a scientist. All science begins with one man's quest for knowledge in the pursuit of a greater goal. That goal may be to develop a product or to coax from seemingly obscure complexity some facet of pure science hitherto not understood. I was just an individual with a quest to seek the truth, to question the laws of physics and the correlations between earth, life, animals, plants, humankind, universe, and the universal truth. Many will say I have lost my mind, but as we progress we understand that science and theory have no limits, and the known and unknown laws of physics create great potentiality to afford all forms of matter that is present in our substance constituting universe to exist.

What I am explaining is not a theory, because I have observed this myself. Specifically origins of our universe can be understood together with the cosmic force, which is the force that creates the cycle of life. The implications and understanding of these discoveries are far beyond what science really knows today. Based on this understanding and armed with this concept, humankind's knowledge will undoubtedly advance one thousand years into a future that was meant to be.

The words in this text may be ignored and find obscurity upon a shelf, or be relegated to a trash can. I am not publishing my findings for notoriety. Rather I am relating my findings in the noble pursuit of knowledge. However when these words travel, they are not only meant for one person who has been selected to dig over these papers with the authority, background, and grandeur of a notable name. It is my sincerest hope that these writings find their way not just to scholars and intellects, but also to people in all walks of life. Let these writings find their way to all professors, physicist, teachers, students, poets, preachers, carpenters, servants, drivers, waiters, waitresses, presidents of companies and presidents of countries, lawyers, judges, the jailed and the jailer, the schooled and the ignorant, the beggar, the homeless, the prostitute, the drug addicted, the drunk, the scorned, the hated, the loved, the forgotten, the ones full of life and the ones that feel life no longer fulfills, the living and the dying, the believers and those who do not believe and every race, creed, color, and age. These words are for every religion, because this is universal knowledge and it is the great equalizer of man and humankind itself.

Humanity, not science, needs to filter out and remove its diluted perceptions and illusions. Our generation will have an awakening of the soul where age does not exist; ego, greed, and selfishness have no place and are a waste of past time; squabbles and feuds are obsolete and irrelevant. Heaven and hell will not exist in our current conventional terms. We will understand that heaven and hell are right here. Heaven and hell, the afterlife, are determined by the choices we make in this life and in every other life that is energy transferred and transformed, no matter which dimension it's in.

Humankind's understanding of reality has been misconstrued and hindered by a few members of the scientific community—not all, just some that are close-minded and others who have a certain degree of tunnel vision. Concepts and formulas will be created that will seem to defy Einstein's theory of relativity, but by the laws of physics itself, they cannot be denied. Many will dispute and some will agree, but in the end the equations will dictate the truth.

Was Einstein Wrong?

For years we believed and understood that there is nothing faster than the speed of light. As a result of Einstein's theory, many scholars believe that there is nothing on Earth faster than the speed of light. While this is generally accepted, a significant number of scientist believe this is not so.

How can we better understand the nature of pure speed of light and the universal phenomena known as the big bang?

We often hear that the first object that is faster than the speed of light is the expansion of the universe. The creation of our universe was faster than the speed of light. Obviously this still does not explain the details.

Einstein's theory of relativity is not wrong; however, it is only right to a certain defined point in time and space. Science explains why objects in space are light years away. When we observe cosmic events, they actually occurred millions of years ago due to the vast amount of space between the place of occurrence and our location. The fact that the occurrence took place eons ago is generally accepted.

This will be true up to that certain point in our solar systems and Milky Way, but at the area of expansion that is occurring now, the future is being created, not the past. The aftermath of the big bang explosion is still occurring within the emptiness of space. Expansion is still happening right now and will continue forever, long after man ceases to exist on this planet.

Scientists also know that time and space have a multitude of properties, but they are presently unable to explain them in terms of how or why. I believe it was because an eternal, formless emptiness was filled. The expansion, formation, and creation of time and space occurred. When the big bang explosion occurred, an unknown barrier was broken—that is, unknown until now.

Many attempts have been made to explain these phenomenal events. None have accurately conveyed any concrete results. The purity of realizing the correlating factors of time and space, along with their functions, will be explained later in detail. These events that occurred are undoubtedly the origins of the universe.

There is more than one way to create an explosion, but there will only be one big bang—for our galaxy, that is. Our journey of existence begins.

Words of Wisdom

All things exist because of emptiness, and to understand this one must understand what emptiness is. There are many factors that complete and compete throughout our existence. Everything, the Sun, Moon, Earth, humans, animals, insects, and every single atom in between have an influence on every other constituent part. To understand these phenomena, you must first understand the significance of two simple words that have been misleading and misinterpreted by many for centuries.

Nothing—the simple word "nothing"—has been misinterpreted for centuries and has been misunderstood by lay people and the scientific community alike. Even many religions use the term "nothing" inappropriately. Even the biblical concept that "the world was created out of nothing" is misleading.

"Nothing" can be defined as a lack of or the absence of a certain factor or perceptible factor. For instance dismissing the importance of something said or thought by saying, "It's nothing," infers that it is trivial. Also, quality, or a lack of quality: "it's nothing special." Nothing can be interpreted many ways. Many scholars, theologians, and educational shows also use the term "it was created out of nothing." This notion is used irresponsibly. Scientifically when defining the universe and its creation, to put it quite simply and directly, the term "nothing" should not be used.

The other word is "emptiness."

Emptiness, in our normal, conventional terms, can also mean lack of something, such as the absence of water in a bottle. The act of removing something tangible from any three-dimensional space is "emptying." To understand emptiness in science, and more so in physics, one must realize what true emptiness is.

Emptiness is an absolute and complete void: no protons, no neutrons, no atoms, etc. It is completely and absolutely void of particles both known and unknown to humanity. Emptiness is a shapeless vacuum that surrounds our universe. This is the true meaning of Emptiness.

This concept can be difficult to understand and also a bit confusing. Therefore I will give a simple analogy. When a jar has nothing in it, we consider it to be empty in conventional terms. The jar is empty of its normal contents, but it is not truly empty because it still contains atoms, protons, neutrons, neutrinos, oxygen, hydrogen, and every other unseen piece of matter that surrounds us and goes through us. The term empty is conventionally used, but the fact is that just because there is nothing in the jar, it does not mean it is empty and void of all things.

In science, like in the physical makeup of our universe, "emptiness" is the supreme ruler of all. The galaxy, our world, every other planet, every star, every atom, every single seen and unseen molecule of anything that exists was not created out of nothingness. That is impossible. These things were brought forth by the absolute phenomena of emptiness.

Emptiness is the defining point of cosmic phenomenon that determines all that is and all that will be. Without emptiness, just an absolute and unimaginable rift will exist: there can be no space for reaction to take place or conformity within the chemical elements to transform into what is necessary at the time of occurrence. A complete void of all elements, a natural vacuum of existence in that space is absolute and necessary for all phenomena to take place. Emptiness is the one non-element that will define and dictate a puzzling universe. Emptiness creates potentiality. But what fills emptiness?

Dark Matter Defined

Now with the understanding of what true emptiness is, one can better understand when I explain the process. In the emptiness of space, from the furthest regions and the outermost edges of our universe and all galaxies, where dark matter of emptiness awaits, not even time exists. Emptiness is formless until the proper conditions exist; it is then transformed into an orb of energized matter and is no longer empty and void. The reason the orb forms structure is simple and will be easily understood shortly.

Formatter of Dark Matter

Emptiness is formless dark matter, and dark matter itself cannot escape it. Once this emptiness materializes into energized orbs, it is unstoppable. When emptiness materializes and transforms itself into Orbs, otherwise known as Dark Energy forms of matter, it is a thousand times faster than the speed of light. Light itself has yet to exist in this area and is, in itself, light years away. Light must wait to amass, grow, ignite and catch up to this emptiness.

The force that creates light, energized orbs of matter must achieve a greater speed than light, because it is the carrier force. Light does travel in a straight path; however the light that is ignited is traveling within a larger orb of energized matter. Space of energized matter must surround it because light cannot penetrate dark matter alone.

Once the process is complete, over time fusion will occur, cause sparks and light will be emitted as a byproduct of the orbs. This creates a halo effect, which completes the process of transformation from Dark Matter into Dark Energy. An explosion is generated, creating light, which races outwards.

Cosmic Attraction

Two known cosmic forces blend together and will play the most dramatic role. Amazingly they are given little attention and, consequently, are not fully researched. They are found on Earth and everywhere in the universe: the cosmic Northern and Southern polarities. These forces dictate absolutely all forms of energy and create many different magnetic fluctuations in the life of the universe.

Magnetism is a key factor in determining the outcome of all other elements because it is the binding force of all matter. How can magnetism have such a dominant role?

Where the cosmic forces of Northern and Southern polarities are naturally created and assembled, magnetic attraction will fill the orbs of emptiness. For magnetism to exist, both North and South cosmic forces must be present. One polarity alone does not define itself. However what I like to call a "free roaming extreme subatomic negative polarity lepton" can create a subatomic positive cosmic polarity force—not to be misunderstood as creating an opposing polarity itself.

The misconception is that the negative lepton dies or decays, but it does not. It is free roaming. The other misconception is that the force is weak, but a single lepton will only be as strong as its outer governing structured force, such as the Earth's magnetic field. The strength of a single lepton is relatively weak, as it should be, because leptons work together. This concentration of leptons creates a greater force of magnetism. If the leptons were in a stronger force at the moments of organization and conformity, a more dense state would occur and deform or crush particles, such as the fluctuating patterns of carbon. Another hypothetical is if leptons were any degree stronger in force, the entire evolution of the life system would be completely different. We would be relatively shorter and the human body's composition would be much more compressed. All of our organs and bone structures would be completely reshaped and much denser. This has been proven in space, where less magnetic force creates weightlessness, and the body gradually elongates to adjust to its surrounding force.

Therefore one of two things will occur. The negative lepton will continue to travel at a highly unrealized speed, because it is faster than the speed of light and will, in a sense, vanish or appear to decay. Or leptons can join with another opposing lepton, creating a larger particle. This, however, is only the first step in creating a bondable magnetic force to fill an orb.

The next stage is a third negative lepton joining in. This will complete and stabilize the two polarity forces and balance out a neutral conducting zone to create an infant orb. The leptons will continue to be free roaming but in a structured pattern. As one lepton escapes another replacement lepton of equal polarity is drawn in. This constant roaming and exchanging of the leptons will occur with each and every lepton at a speed greater than the speed of light. This creates the force of dark energy or electromagnetic forces.

My definition of this process is what will be known as the cycling process of leptons and is, in fact, what gives magnetism its orbital motion patterns. Once this process is formed, magnetism can increase its force in the cycling stage. If the third lepton is not joined with the other two, the two will separate and disperse back through the cosmos.

Orbs, the Creation of Dark Energy

The first particles that fill this empty orb will be and will always remain the cosmic Northern and Southern polarities. This unseen, most extreme subatomic negative cosmic particle creates magnetism.

Once joined together magnetism travels in an orbital structure and in an orbital path. Therefore this will define the structure of Emptiness as an Orb of energized matter, as stated earlier. This process will create the halo effect of what is conventionally recognized as Dark Energy.

The Creation of Time Itself

With this massive bombardment of North and South polarities, another phenomenon takes place. Where there was once emptiness, time and space have their start and a degree of accountability. Time and space is actually formed. Time cannot exist where space does not exist. As the emptiness is filled, the Dark Matter that surrounds our universe is slightly pulled in by energized orbs, filling and energizing the Dark Matter. Then it is pushed away. As a result space is being created and our universe is expanding. That expansion is faster than the speed of light.

Then and only then can time begin to exist. Time is at its infancy. As the progression occurs and that specific space is filled and passed, the time and space continuum shall increase. Therefore time and space are now relative in terms. All future events are at hand, doubling at the speed of light.

What Does This Mean For Us?

If we can understand the relationship between and the creation of time and space, then we will understand everything else. Every question beyond our wildest dreams will be answered. The reason magnetism is the governing influence on our universe is that aside from determining the outcome for every element, magnetism seems to explain other phenomena that occur in our universe.

On an earthly scale, other reactions and dynamics are byproducts of this massive Northern and Southern cosmic force. The violent releases of these magnetic forces are in the form of a spark due to the breaking, escaping, and dispersing through this orb. The byproduct of the release is various manifestations of energy, sometimes also seen as light. This dispersion of forces and polarities must aggregate in extraordinarily large quantities and in equal amounts for the ignition to spark the creation of energy.

Actual Night time Unaltered Photo Experiments

Dark Matter and Dark Energy

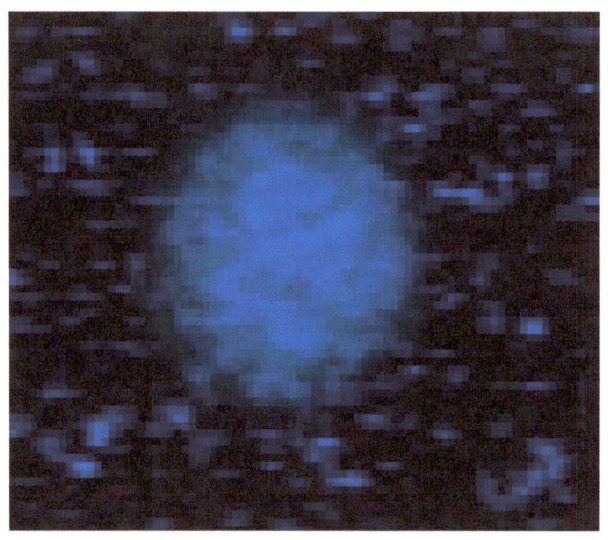

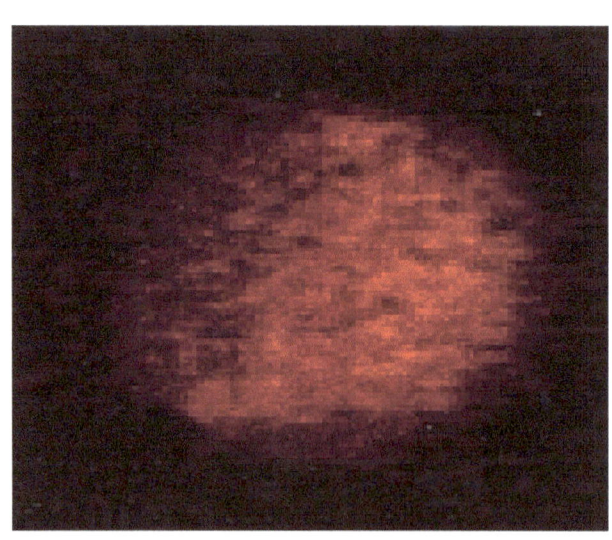

The orbs are formatted Dark Matter and Dark Energy.
Illustrated are different stages of the cycling process.

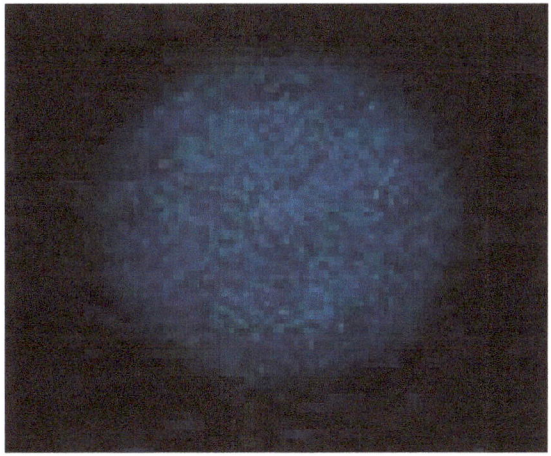

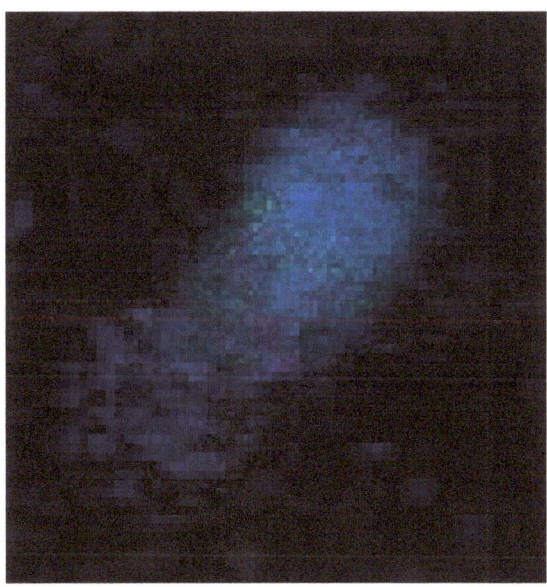

A dispersion of Northern and Southern polarities are creating an energy burst into a Vortex (possibly the first photograph of an actual Wormhole). This event is several light years away and many miles wide.

All photographs depicted are unaltered, magnified and were taken with a polarized lens by G. Davila.

Magnetism vs. Gravity

Magnetism and gravity have much in common. Gravity is believed to be one of the first forces to develop immediately following the big bang. However I believe it has been misunderstood and misinterpreted. Magnetism would undoubtedly have to have been the first byproduct of this mass explosion. Dark Energy was escaping and creating fluctuations of formatted particles. However this magnetism will be distributed at different wavelengths and polarity strengths due to the distance of projections of energized matter, pushing outward the surrounding dark matter. This creates a force conventionally known as gravity, which can now be defined as a specific magnetized level of Dark Energy. Cosmic magnetism levels would have been constituted in fluctuating strengths between each ripple of the shockwave pattern after the big bang explosion.

Magnetism and gravity travels at the exact same speed of about 9.807 meters squared per second. Nonetheless a different formula is used. Objects fall at the same speed in the same gravitational environment or field. However one can also state that things are being pulled down by Earth's magnetism at a predetermined force of polarity or in relation to the vector force of polarity.

Powers of Magnetism

Magnetism and light also have much in common. As previously explained, orbs of magnetism emit a spark when there is a mass bombardment of polarities. The byproduct is a form of spark/light reaction, just like stars and every other static generated form of energy power or nuclear power that is created.

The spark that is ignited can also be extracted to form its own byproducts such as protons, neutrons, and electrons, the make-up structure of an atom, to create the dynamic force of energy. Commonly this is known as electricity, but conceptually it should be called magneticity.

The nomenclature is not important at this point of the discussion. Understanding is what matters. This is how electricity is made. Without magnetism there is no electricity, period.

Magnetism takes many forms. Gravity is magnetism, magnetism is gravity. Magnetism creates electro-magnetism, electro-magnetism creates magneticity/electricity, magneticity/electricity creates electric currents, electric currents create power, power creates light. Magnetism even creates other cosmic rays, radiation, and sound waves that are known to us. Once understood mankind can reproduce these forces.

Absolutely everything that is created is held together by emptiness of dark matter and the magnetism of energized matter. This allows for reactions and conformity to take place and then be held together by polarity amalgamation process. Amazingly as much as magnetism is used and transformed to energy, it is never depleted. Northern and Southern roaming forces will be redistributed to the cosmos and, in a true sense, recycled, and be transformed from one energy form to another.

Splitting the Atom

Let's take an atom for example: it is supposedly the smallest particle with unique identifiable properties known to humankind. Just because we cannot see an atom does not mean it is not there. Using scientific methods we can experiment and determine its existence. There are several questions you must now ask yourself: what force is holding the atom itself together? If the particles of an atom are attracted to one another, what are the forces that attract or bind them?

One can also notice how they are traveling in an orbital structure as well as in an orbital path, a path seen before in the cycling process of leptons into magnetism.

Accordingly one must conclude that not only is magnetism in some way equivalent to the speed of gravity, but it is also equal to the speed of light, emission of electricity, and other byproducts. Moreover what we know as gravity, speed, light, and energy are all relative to one another because they are, in fact, the same.

This is much to comprehend, and one must have an open mind to entertain the concept. But when you investigate the facts and formulas, you too will come to realize the truth and, perhaps, also have a life changing epiphany. This is notwithstanding that its factual basis has been evident.

Time to Go Nuclear

Every elementary student knows that every star is a sun and every sun is a star. It is only because of its proximity that our Sun was given a conventional name. This correlation will determine the shape and formation of our galaxy. The universe will be dependent upon the placement of these stars and planets. Each and every star will act as magnetic space (energized matter) stabilizer zones. These stars will define our galaxy's shape during this expansion.

Now let's put all this into perspective. Imagine the largest concentration of an energized orb you can fathom, full of Northern and Southern magnetism. Let's say about 1,299,400 times the size of the Earth. Both polarities are breaking out of its orbital structure at such an incomprehensible rate that heat it generated. It creates a nuclear byproduct, emitting forces, polarities, fluctuations of magnetism, plasma, gamma, X-rays, and beta frequencies from every sector of the electromagnetic spectrum. This could create a massive ball of energy and light similar to the Sun. What if the cosmic force that fuels the Sun is the fusion process of magnetism itself?

It sounds impossible, but we all know light travels at about 186,000 miles per second. We also know that the distance from the Sun to the Earth is approximately 150 million kilometers or 92,935,700 miles. We must also take into account that due to the Earth's shape and rotation, our actual distance from objects in outer space differs by several hundred miles from point to point. Now take the formula I am putting forward. If you take the speed of magnetism on Earth (which is coincidently exactly the same speed as gravity, 9.807 meters per second squared) and multiply it by the distance between the Sun and Earth, what would be the answer? Better yet employ the formula yourself. Look up the speeds and distances from an independent source yourself. Your total will equal approximately 186,000 miles per second. This is not ironic. It is one principle I am propounding.

Breaking Einstein's Theory

This is where Einstein's theory is correct with respect to the speed of light. Nevertheless as far as the expansion and the creation of light and the universe, it is incorrect. From the time of Einstein's equation until now, the speed and expansion of the dynamic universe would have had to increase at a rate of 9.807 meters squared per second, along with the accompanying speed of orbital force that pushes dark matter, creates Dark Energy and the light itself.

It is believed the fastest speed is 186,000 miles per second. But this is also misleading, because once you pass Earth's magnetic field into a certain point in space, also passing the Sun's magnetic force, the distance is multiplied squared. This would again explain the expansion of the universe at an unbelievable rate of speed. This is where according to Einstein's own theory, the forces would come into effect: the greater the energy the greater force into a greater mass. The more dominant force would regulate the lesser force.

It is a complete certainty that cosmic forces that regulate the Sun and created the Sun are emptiness (Dark Matter) along with the Northern and Southern magnetic polarities (which create energized matter). Another factor that can be taken into account is the fact that the Earth is rotating around the Sun's magnetic force. As previously stated magnetic force is dependent on the size and distance of the magnetic object. On Earth we are regulated by the Earth's magnetic field, which encompasses the distance of outer space up to the Sun's magnetic force. Nonetheless the greater magnetic mass is the Sun, and it defines Earth and its patterns.

If the Sun defines Earth's path, what defines the Sun's path? Our entire galaxy is orbiting in a spiral path because of the polarity of the Milky Way. Why?

It is generally accepted that Earth is 3.8 to 4.5 billion years old. The Sun is a second-generation star that is approximately 4.6 billion years old. Our universe is thought to be between 13.5 and 14 billion years old. The Milky Way is between 200 and 300 million years old. The universal understanding is that the big bang created our galaxy and Earth. Can we find such an event in our own galaxy? Yes we can. Even in space every explosion must leave a trace, such as a deformity or hole, like the one in the Milky Way, which has at its center a deformity: a beautiful black hole.

Origins of the Universe

How probable is it that the big bang and the center of the Milky Way's black hole are the remnants of one event? Every explosion, no matter how small or large, leaves a trace and a specific pattern; such as a gaping hole, like the one in the Milky Way. This means the big bang and beginning-less time have a starting point for us. The center of our universe, the Milky Way, and its mysterious core (the black hole) are the true beginnings of our universe. The cosmic event known as the big bang that started our universe that we know today.

As explained earlier, when there is a massive buildup of magnetic polarities, a speed of orbital magnetism, and the rate of force of that magnetism in that particular orb, the North and South polarities will become concentrated and grow larger and larger. The orb and surrounding magnetism is dependent on the size and polarity of that force. As the dark matter is transformed into Dark Energy at an uncontrollable speed, nuclear fusion will take place and a massive explosion will occur. This is where magnetism will demonstrate the largest explosion possible: the big bang.

This phenomenal event would explain what is enabling the galaxies to expand in a universal blast pattern. The greater the force in magnetism, the greater the orbital pull and push effect. The larger the structured pattern, the larger our universe will get.

The Space of Emptiness

We can conclude this because it is understood that such defining factors as emptiness (or dark matter) and an unknown force were generated. Unknown forces created a nucleus of particles, and then a massive increase of heat began a fusion process, which triggered a nuclear-type explosion. The unknown generating force must be the polarities as previously explained, because they are the first to enter matter and transform it into dark energy. Massive uncontrollable buildups of magnetic forces are produced. The cycling process of leptons begins to form into an energized orb. The swallowing of space/dark matter occurs, resulting in an explosion due to its uncontrollable speed and buildup. This will create a big bang and leave its mark: a black hole. This reaction was possible because there was no other Dark Energy force in the immediate space where the event occurred.

When Stars Collide

We know most stars in our galaxy are remnants of older stars. Stars that have finally met their life span can either dissipate or go out with an explosion. Some are also believed to create black holes.

Stars themselves are of different masses. To a certain degree, we can determine their life span. We understand that when a star is dying, it begins to lose its velocity. Yet the potential energy still exists in that star.

Let's find a double star system in our early galaxy. For a simple example, let's call the two stars Mega and Minor. The star's mass would determine the life existence of each star. What if the mass of one star was substantially larger than the other? This is not far-fetched, because we can see this in the Alpha Centauri system.

Once Minor's life comes to an end, the kinetic energy that was present still has orbital motion. Its motion will inevitably be reduced almost completely. However as Minor's chemical energy and fusion reaction dissipates, a large amount is still present in the orb. Mega continues on its normal rotation. As time passes, Mega begins to pull, via magnetism, Minor into its magnetic force—that which is surprisingly similar to the black hole effect, the difference being the black hole is no longer a positively acting star. This action will undoubtedly cause a cataclysmic collision.

Once the stars collide, a natural explosion will follow. I believe this unimaginable galactic event releases the potential energy to disrupt matter to form black holes. This event can rip the fabric of time and space. What we perceive as a "hole" is really a dispersion of dark matter and energized matter.

True Nature of Black Holes

Faster than the speed of light, what is a black hole? The understanding of black holes is a rather new and understudied realm. Many answers seem to jump the questions. Not much is known about black holes. We are just beginning to understand how they originate, where they can be formed, and why.

Currently, the general consensus regarding black holes is that they appear to destroy everything. What we need to learn is why and how the black hole acts the way it does.

How does a black hole such as the one in the Milky Way operate? The big bang created life, but it also takes life, or the force of life. Once a black hole is created, it appears to destroy everything in its path. However it is not completely a destroyer. Beginning life as an infant star, even if it was only for one-hundred billionth of a second, it was acting as a positively charged force of energized matter. The black hole was an emitter of positive energies.

An exploding star with a broken orbital structure forms a black hole. The explosion will force a dispersion of dark energy. A black hole is acting as it should be, no longer a positively acting star; it is a negative energy force of dark matter, just like the negative force of leptons' cycling that created the big bang.

The black hole is no destroyer: it is removing every single North and South particle lepton in its path, trying to fill itself and regenerate, but it cannot. The speed is so furious that light appears to be trapped. The "swallowing" of light effect can be seen, but what is actually occurring is the removal of particles. The black hole is taking back its infant forces that it once emitted and gave freely. Other transformed particles are tossed back into orbit to be absorbed and transformed later. The location of the big bang explosion that created our galaxy, eons ago and light years away, has finally been found.

Conclusion

There was only one big bang event, and it is true that there will never be another one of that magnitude. The space in our galaxy is already full of orbs of dark energy, which were apparently not present at the beginning of the original formation of the expansion and big bang explosion.

In my search, I came to one conclusion: how can a simple answer of "yes" positively influence and give explanation to all these questions? After years of research and experiments, I do, in fact, believe these are the answers. Perhaps we will be able to answer more monumentally complex questions about life, Earth, and universal properties. If these are the answers, then much more can potentially be explained.

It is important to understand this process of pressure and pull of dark matter and push of dark energy that flows outward. These factors create our galaxy's conventionally termed "gravity," which is really fluctuations in magnetism.

Emptiness of dark matter, along with Northern and Southern polarities are cycling into energized orbs of dark energy. This cycle process is what creates time and space. In our early universe this process created an infant orb which exploded to form a black hole. The dispersion of matter within the black hole created the vacuum at the center of the Milky Way. This all helps define the rotations and distances between objects. The explosion and potential energy release explain the expanding universe. The origins of the universe and our existence have been located; the once puzzling equation of the Grand Unified Theory has been solved.

An absolute understanding of this will bring us into the realm of astrophysics, the creation and the probability of manipulation of time and space itself—a subject to be discussed at a later time.